Let An Earthworm be Your Garbage Man

THIS report is primarily about a new system of garbage disposal. It tells how, on an all-year-round basis, earthworms and soil micro-organisms can transform garbage and garden wastes and surpluses into useful humus for soil improvement, with a minimum of inconvenience and in surprisingly short time.

Raw vegetable (organic) matter is, as everyone knows, not in the right condition for nourishing plant life. It needs to be composted, the process in which the material is broken down, digested and ultimately transformed, by actions of various descriptions, into the finely divided, black and crumbly stuff that makes all the difference between soil sterility and soil fertility.

The ability of worms and the microscopic life of the soil to process raw organic material into humus that plants and the soil can use has long been known. It has been the basis of various systems of composting, all of which had serious disadvantages. The field research which forms the basis for this publication was undertaken to determine whether these disadvantages somehow could be simply overcome. It was done on an experimental farm in Vermont, where winters are long and cold, because the most serious difficulties arise in winter. Then earthworms are inactive far underground, if alive at all, and miscroscopic life ceases to operate . . . hence composting usually ceases.

The work was undertaken because of two considerations. The first is that garbage disposal is something of a problem in most rural and suburban areas. Even if the community supports a "free" disposal service, the thoughtful homeowner hesitates to avail himself of it. There is just too much good stuff that ought to be used, not wasted. Burning of garbage is wasteful too, and may cause odors that the family and neighbors properly object to. Flushing it through a sink disposal unit is no less wasteful and may cause serious septic tank troubles. Burying it, assuming sufficient land to use for such a purpose, is not only wasteful but very laborious, if you are going to get the stuff down deep enough to avoid the attention of rats, dogs and other hungry animals.

The second consideration is that good compost, the equivalent of well-rotted manure, is highly desirable on any country or suburban place. There is no such place where enough of it is naturally available, unless fairly large scale livestock projects are under way.

Because earthworms are involved in the system we have evolved, there is much attention to them in the report that follows. They are controversial creatures in some ways. Basically, no one questions that they are useful, real friends of man in his efforts to draw the most out of the soil he cultivates. They are aids to composting, and are so considered in this report. Just to what extent their benefits extend, and to what extent special or so-called "hybrid" worms may be useful, is where the controversies begin. We have gone to the highest authorities for views on these points, and the reader must be his own judge of what to believe where the more extreme claims are concerned.

There is some attention to the profit possibilities in earthworm culture, both through sale of the worms and their castings, often used for potting plants. We also consider the value of "activators" which are added to the compost material to speed up the process.

How things grow without worms — and with worms. See Page 36. (U. S. Dept. of Agriculture Photo)

TRIAL and (mostly) ERROR

SHOWN above is the type of compost box that a while back was considered the real answer to how to compost garbage and garden material. The box was located at the homestead of Ed Robinson, widely known as the author, with Carolyn Robinson, of the *"Have-More" Plan.*

The idea was to fill in one side with raw material as it accumulated, adding soil from time to time. Then after filling one side and allowing the composting to go on for a while, all was turned out and mixed into the other side. This was quite a chore.

However, the real objection to this system was not so much the work involved as the messy and smelly character of the setup, plus the accessibility of the material to any kind of varmint, including especially rodents, and to flies — a real breeding place for objectionable insects.

Furthermore, the compost box is useful only in mild weather. As we noted, warm weather is needed for the kind of biological action which brings about composting. There is no practical way with such a box to maintain favorable conditions in freezing weather.

The box was an improvement over an even older scheme, which was the compost pile. This required pretty careful organization; alternate layers of manure, refuse, etc., and soil, with sprinklings of other materials. The whole had to be kept as straight-sided as possible; not heaped up into a pyramid which is the way a pile ordinarily accumulates. The pile had all the objections found with the box, as well as being a little more difficult to work and somewhat less efficient in action.

With the experience of Ed Robinson and others as a starter, various improvements were tried, with the objective of eliminating or reducing the drawbacks experienced with the piles and boxes, and making the whole process more expeditious and efficient.

As in so many instances, progress came with trial and error.

A shallow hole for garbage, spaded out as above, has some definite advantages. Buried under soil, the material will be composted with astonishing speed, and will be available at the point of burial for soil improvement and nourishment of whatever is planted in the area. Theoretically, you can bury the stuff and forget it.

But the varmints won't let it alone. They will burrow into the ground, unearth the garbage, spread around what they don't want, and in general multiply undesirably.

The heavy boards shown in the illustration were placed over the area in which garbage had been buried, to keep dogs, cats, skunks, rats and so on from excavating the stuff while it was still nutritious for animal life.

This did not work. The animals and rodents went right under the sides of the boards, no matter how broad or how heavy the cover was laid down. Also, burial presents additional problems in muddy weather or winter when the ground is frozen solid and often covered with snow and ice.

Even simpler than spading a hole, as an initial operation, is digging a furrow with a Rototiller. The garbage is distributed along rows by this system, and in two months, if the soil is warm, is sufficiently composted to enable planting of a crop.

But the same disadvantages hold for this as for the digging of a shallow hole or pit. Pests dig up the stuff, and boards laid along the furrow do not solve the problem.

It looked as if the logical answer to the difficulties that kept cropping up was a pit of material that animals could not penetrate.

Accordingly, as above, a cement block pit was built on top of the ground.

The blocks form a square and do not cover ground inside, as this would cut contact of the material to be composted with the soil — undesirable because from the soil beneath come the worms and other organic life to help with the composting job. This also did not work, because the varmints would burrow under the blocks that formed the walls of the pit.

FOUND — A Way That Works

Here is a step by step account of the simple system that was worked out for garbage composting without the difficulties that had been experienced in the other methods.

First step — dig a hole the depth of one concrete block — eight inches, in other words. Also three blocks square, or four feet each way. Pick a spot — any spot — in or near the garden or wherever else you will want to use the compost. Only caution is — don't build *Winter* pits where you will have trouble getting to them through snow or mud.

Second step — here the first course of blocks has been laid down. The blocks cost about 20 cents each delivered, and of course can be reused indefinitely. Cinder blocks are lighter and just as good for the purpose.

Three pits are ample to handle all the garbage available from a family of four. Locations may be reused, or changed. Then the old pits are filled in with topsoil or sod scraped from any handy nearby location.

Step 3 is to add another course of blocks. Then you are ready for Step 4 which is the actual beginning of composting — by starting to fill the pit with garbage. The illustration shows the pile of dirt dug from the hole, in right foreground, and cans in which garbage has been collected, on the Countryman's Cart at the left.

Note this handy vehicle. It has no end of uses around a country or suburban place, and will transport heavy objects with much less work for the user than a wheelbarrow, besides having a greater capacity. The big wheels will go through mud or snow, and over rough ground with little effort. It is extremely strong and durable. For more information about the Countryman's Cart write Home, Farm and Garden Research, Inc., Box E, Noroton, Connecticut. Ask for circular and prices.

The two cans shown are big enough to collect the garbage that arises in two to four weeks, depending on size of family and season. The filling of the pit can be donc prctty much at your own convenience, and in favorable weather.

Here garbage from the two cans had been dumped in and leveled — note that it comes less than half way up the bottom course of blocks. Anything of organic nature goes in, even such resistant materials as citrus rinds and bones (all except the heaviest beef bones will break down in time). You exclude, of course, glass, cans, etc., but light paper bags cause no serious problems. Even your vacuum cleaner dust can go in.

Here's Step Five, covering the first layer of garbage with two or three inches of soil. This inoculates the pit with earthworms and soil micro-organisms — some will come up from under also. The soil layer kills odors and avoids attracting flies. The whole job is clean as well as efficient — your compost will look and smell like good rich earth, rather than the product of decay. Composting is *not* decay if done right.

Step 6 — To conserve moisture, which is so important to the activities of earthworms and soil micro-organisms, cover the garbage-soil layer in the pit with two burlap bags. Do not cover with any material such as sheet metal which would exclude air — the worms, microbes, etc., need air. Part of the value of the earthworms is that they keep the pit aerated, by tunnelling up and down through the heap.

An optional step (let's call it "6A") is watering the pit in dry weather. This insures that the composting action will not stop; otherwise it is unnecessary. Of course, if you do not add water and do have dry weather, the composting may take longer than the three months which we have found to be about the time needed after the pit is filled. Then you might need more than three pits.

Step 7 — After putting down a layer of garbage and soil, and covering with burlap, lay down a wire cover — 1″ x 2″ mesh "turkey wire," available at most hardware stores. Weigh it down with whole or half blocks as shown.

With this setup, *nothing* gets in except the air and the rain. Insects are not attracted, animals can't penetrate the defenses.

You continue to build up the pit for about three months. This will probably involve laying two, and perhaps three, more courses of blocks. Then leave the pit for three months and you will have your compost. It is as simple as that; there is no turning of the pile needed, no moving of the material from one pit to another.

When time comes to use, the compost can be easily shovelled out by simply setting aside the blocks from one end of pit. If compost is to be used in pots or flats you may want to sift out the larger bones which have not as yet completely disintergrated. This is not necessary if compost is to be used directly in garden.

It Works in Winter Too — Here's How

To carry us through the long Vermont winter, we dug two pits. Naturally this had to be done before the ground froze. Because we expected plenty of need for dirt, to build the pits four or five courses deep, we dug twice the warm weather depth — 16 inches for two courses of blocks underground.

We were not concerned because the deep-digging meant bringing up some sub-soil. The addition of the organic matter, plus the mixing and grinding activities of the earthworms, is going to turn that subsoil into prime topsoil. This is a real bonus value! It is particularly valuable in areas where the subsoil is high in minerals.

To keep the dirt from freezing, we hauled some hay, which was available on the farm. Not the best hay, but the worst, some that had been spoiled by poor weather at haying time.

The hay was applied thickly to cover the dirt for maximum protection. It is easily pushed aside when the time comes for digging dirt to throw over the stuff in the pits.

Here's how the pits looked waiting for winter. Three courses deep, extra blocks parked handy. The pit in the background is from earlier in the year. Material for covering the dirt in winter may be almost any kind of hay, including marsh hay, or any kind of straw, or sawdust, or the chicken litter materials available in many areas.

In rural areas it is easy to get straw or hay for as little as 75 cents a bale. In suburban areas the prices will probably be higher. Sawdust is very cheap or available for the cost of hauling it, but somewhat harder to handle.

Whatever is acquired as cover for the dirt is reusable. It will eventually break down and merge with the soil, acting for soil improvement — so your investment is returned in two ways: maintaining the process in the winter and supplying additional organic matter.

Garbage cans kept in the barn in below zero weather will freeze. (In a barn in which cattle are kept they would not freeze). We took them into the kitchen for a couple of hours to thaw the garbage just enough to loosen it from the sides of the cans — not all the way through. This is a little trouble, but there is no odor.

Here is real winter! Plenty of snow and still falling. A night before low of 21 below and just under zero frequently. This view was taken in January. Note how the big wheels of the Countryman's Cart easily negotiate deep snow.

Snow collects, of course, in the pits, although it tends to melt there sooner, because of higher temperatures well below the surface, and protection from winds. Here snow is a foot or more deep.

We do not bother to take out the snow; just dump the garbage on top of it. Then we continue as in summer with dirt and wire. There is no halt in any of the process. Burlap covering may be left out during winter months.

The dirt under the hay has not frozen solid. It's usable all winter if you keep it covered after you've got what you need for a layer of garbage.

This ends the account of how to handle the garbage pits. While they are being filled, and thereafter until the composting process is completed, action goes on, even in winter. It is a fact that composting generates a certain amount of heat, and prevents solid freezing.

It was noted that about three months is the usual time for completing the making of humus, during most of the year. Somewhat more time is required in winter, because although action does not stop, it slows up. This has no practical disadvantage, however, since you are not ready to use the material until the soil is warmed up and ready for working.

When it is time to prepare a garden seedbed, the first of the pits you prepared for the winter's garbage should be completely composted, ready to yield a rich black hoard of finely crumbled material, free of undigested vegetable matter, sweet-smelling and easy to handle.

The account of how to do the job has left many questions unanswered. Before getting into the fascinating story of how earthworms and soil micro-organisms operate on garbage, let us first consider some additional practical matters.

Adding wood ashes. They furnish potash particularly, but also other plant nutrients and trace elements, thus enriching the compost. Hardwood ashes are to be preferred over softwood ashes, but any wood ashes are beneficial.

SHOULD YOU USE "ACTIVATORS"?

IN the effort to make our system as simple and inexpensive as possible, we for the most part avoided the use of any activators or special earthworms. We found we got along satisfactorily without them. But there is reason to believe that some additives would be valuable either for accelerating the composting action or making the compost a complete fertilizer for garden use, which it is not, as made without supplements.

The reasons for this will appear in more detail later, but essentially the facts are these: earthworms and soil microorganisms have food needs which are basically those of most living things. They need mostly nitrogenous food (food nitrogen is protein) and carbohydrates. The latter they get in abundance from the organic matter. The former is lacking in sufficient quantity in most garbage accumulations — sufficient, that is, to maintain maximum life and growth for the organisms. An exception might be where there was a large proportion of meat scraps.

If the worms and micro-organisms don't find enough nitrogen, they have to wait until their own processes form the necessary element. So there is sound reason for adding nitrogen in some form.

The most convenient form for adding nitrogen is through a complete fertilizer. This also adds phosphorous, which the organisms also use, potash, needed for the soil, and a number of trace elements which are carried by the superphosphate in the fertilizer. Also these are useful and will not only speed up composting but later eliminate the need for use of fertilizer in the garden, provided enough of the compost is spread. *Such a fertilizing mix can never "burn" your plantings.*

Only average recommendations can be given. The nature of the material composted varies, and it is not practical to attempt to adjust quantities to such variations. A good and safe amount to add is one-half pound of say a 5-10-10 analysis fertilizer to one cubic foot of material — soil and garbage — in the pit. No need to measure accurately, just sprinkle it on occasionally, with the objective of using up 8 pounds of fertilizer by the time the pit is filled up to three courses deep.

If 5-10-10 isn't available (almost any farmer's supply house will have it but not the average seed or hardware store) use 5-10-5 or 4-8-4, the analyses of the commonly sold garden fertilizers. The idea of the 5-10-10 is to bring the analysis of your compost up to about an even ratio between the three essential elements. Such a ratio, expressed commercially in analyses of 8-8-8, 10-10-10 or even higher, is increasingly being recommended by the experiment stations for general crop nourishment.

Why does a 5-10-10 fertilizer end in a compost with equal proportions of the three essential elements — twice the nitrogen that comes out of the fertilizer bag? Because your compost has given you this wonderful dividend in available nitrogen.

Some people add lime to compost. This is a dubious practice, recommended if at all because at times and temporarily there may be an excess of acid in the material during composting. Rather than incur the risk of an undesirable alkaline condition, bad for composting action, it is better to avoid lime, except for what you get from the eggshells and bones in the garbage.

There is another type of additive which is likely to come into greater prominence, with more knowledge and experience. This operates either through adding useful organisms or stimulation for such organisms, or both.

The British have done much work in such matters and doses for compost pile or pit which seem at first hearing to be eccentric if not humorous are now being taken more seriously. A product that enjoys wide favor for activation purposes is described as composed of "certain herbs and honey." The herbs are believed to harbor organisms, or produce hormones, which are effective on compost, and the honey would supply nourishment to keep the organisms active.

Elsewhere in this publication we refer to the work of Dr. Ehrenfried E. Pfeiffer, who with microorganic additives composts garbage commercially faster than it ever has been done before. If and as his or similar discoveries are applied to conditions which can be maintained practically on a small scale, some major speedup in composting action may become feasible.

As to the addition of worms: we have never found this necessary. It may be that the manure worms which are sold widely have a faster action on the garbage than the common field worms. This is a question that will be gone into in some detail. Also, our good clay soil in the Champlain Valley is more likely to be plentifully supplied with worms than a poor or sandy soil. Adjustments to local conditions must be made.

What To Do With The Compost

IT is easy to tell when your compost is ready for taking from the pit and using in the garden. It is dark and crumbly, resembling earth that is light in weight but very dark, nearly lacking any obviously recognizable parts of the original materials.

It is ready for soil improvement, the equivalent of well-rotted manure.

Why is "well-rotted" manure customarily recommended rather than the fresh products? One answer is that the fresh manure may "burn" the seeds or plants, and defeat its purpose of aiding in growth. Although this may be true of the so-called "hot" manures, the compelling reason for not using fresh manure when planting is to be done immediately lies in another consideration which we have already touched upon.

In its fresh state, manure or other organic substance, such as the garbage we have been busily disposing of, is in the soil at once attacked by the soil microorganisms and earthworms, because it is high in the carbohydrates they like to feed upon. In doing so, as we have seen, they will use nitrogen also, and this they will obtain mostly from the surrounding soil. Net result: nitrogen deficiency in the soil until the raw material is rotted into humus, a more resistant form lower in carbon, which will continue to be decomposed — for the benefit of plant-life — but at a slower rate and with other processes which release rather than use up nitrogen.

Organic matter is necessary for the growth of plants in the soil. With it in abundance, soil is really fruitful. Organic matter improves the soil by increasing its water-holding capacity, keeping it open to the entry of oxygen which the roots need, and contributing plant nutrients, among other benefits.

Added to compact clay soils, it lightens them, makes them more friable or crumbly. It closes the pores of sandy soils, and supplies binder material — a natural soil-conditioner — for giving a crumb structure to the earth.

As organic matter decomposes under action of the soil, a continuous supply of carbon dioxide, nitrogen and phosphorus compounds are released for plant growth.

Organic material "buffers" the soil, preventing excesses of acidity and alkalinity. It helps to hold nitrogen compounds, calcium, magnesium and trace elements against loss in the excess water which drains down through the soil.

There are benefits enough to repay any effort in making compost and applying it to the garden. As to the latter, one good way is shown on this page — working it into the garden or flower bed with hand tools such as rake and hoe or the electric Rototiller.

This electric Rototiller is an amazing new power tool. With its various "Snap-On-And-Off" attachments it digs and hoes, edges and trims lawns, cuts wood and mixes cement. It also is a super-power electric drill — and even shovels snow! For more information and prices write Rototiller, Inc., Dept. E, Troy, N. Y., and ask for literature on the Model E Rototiller.

Larger garden machinery, such as the Rototiller above, will readily incorporate large quantities of your compost into the upper five or six inches of the soil — best not to bury it too deep, as its main benefits are exercised near the surface.

Compost has important benefits over raw vegetable matter such as cover crops, for soil improvement. It can be thoroughly mixed with the soil, and it does not have to go through a preliminary breaking down before releasing nutrients.

Here's a strip gardening idea which enables the compost to be concentrated where growing is actually done. Leroy T. Stratton of Rutland, Vt., shown at work, keeps paths between the rows of his garden in grass, which he mows rather than cultivates. Paths are better walking, especially in wet, muddy weather, and the green bands are beautiful to behold.

All fertilizer and compost are worked into the growing beds. It's not only about the best looking but one of the most luxuriant gardens we've ever seen.

One of the handiest uses for your compost is for starting seeds in flats or pots, and, of course, for potting house plants. Here are shown 100 Vita-Pots filled with compost early in the season. Tomatoes, cabbage, broccoli, and such, even melons, cucumbers and corn can be started this way under "cloches" as shown below.

"Cloches or "PMGs" (Portable Miniature Greenhouses as they are called) are not as yet widely known in this country, but are used by the millions in England and Europe. They are a great improvement for cold frame work, as shown here, with or without a heating cable, for early starting in flats, pots, or right in the garden itself. They protect against frost, wind, animals and birds. For more information and prices, write to Waldor Greenhouses, Dept. E, 150 Washington St., Salem, Mass.

The Facts About Earthworms

The following section of this report is contributed by Henry Hopp of the United States Department of Agriculture and without doubt the world's leading authority on earthworms. The department set him to study earthworms because of the many inquiries they were getting about the characteristics and values of these creatures. He digested all the previous work in this field and then embarked on some experiments of his own.

We went to Dr. Hopp to get an authoritative account in terms of the layman on the work he and others have done. The facts about worms are important for anyone who does composting or wishes to have better soil for farm or garden. You will find that his studies have made Dr. Hopp an enthusiast about earthworms, albeit with a scientist's caution about claims that cannot be experimentally proved. In reading his account it should be kept in mind that what the worms do in the soil is essentially what they also do in the compost pit. You will wish to encourage the maximum earthworm population in either after reading Dr. Hopp's absorbing report.

MOST every farmer and gardener is aware that earthworms occur widely in agricultural soil. They are especially frequent in the richer soils, although they also occur in ordinary garden or farmland. The association of earthworms with productive soil causes people to wonder if earthworms play a positive role in soil productivity or merely prefer to live in the better soils without contributing to their productivity.

This is not just an academic question. It is one that can vitally affect the productivity of many millions of acres of agricultural soil where farming or gardening is being carried on without regard to what it may be doing to the earthworm population.

There are over 3,000 species of earthworms in the world, but only a very few that are important in the tillable soils of this country. Some common ones are shown in the accompanying illustration (Figure 1).

The night crawler, *Lumbricus terrestris,* is the largest of our earthworms. It is more common in the northern states. Heavy fertilization seems to favor its development in meadows and lawns.

Helodrilus caliginosus, with its variant known as the form *trapezoides,* is described as the common field worm. It occurs throughout the humid area of the country. It is more common than the night crawler particularly in the southern states. In the same locality, this species may prevail where the fertility level is too low for the night crawler.

On soils of extremely low fertility, neither of these species prosper. In rundown bromesedge fields around Washington, D. C., for example, the main kind is the small slim worm, *Diplocardia verrucosa.* It has no English name. There may be quite a few of them in the soil, but its holes and casts are so small that it has only a minor effect on soil properties.

Another species found quite widely in agricultural soils is the green worm, *Helodrilus chloroticus.* It is a rather short but stout worm of typical greenish color. Only because of its prevalence is this worm deserving of comment; it is actually quite inactive. Very often it is found curled up in a semi-dormant condition while the other worms are active.

Two other kinds worth mentioning are especially common in compost piles. One is the so-called manure worm, *Helodrilus foetidus.* It is known also as the brandling or red wriggler, the latter because of its squirming reactions when handled. This species can be told by the transverse rings of yellow and maroon which alternate the length of its body. The other is the stouter, *Lumbricus rubellus.* It is a deep maroon color and does not have the yellow bands of the manure worm. Both these earthworms invade refuse, although the former is more prevalent in manure piles. Neither occurs commonly in agricultural land; although they will come in where large amounts of refuse are added to the soil.

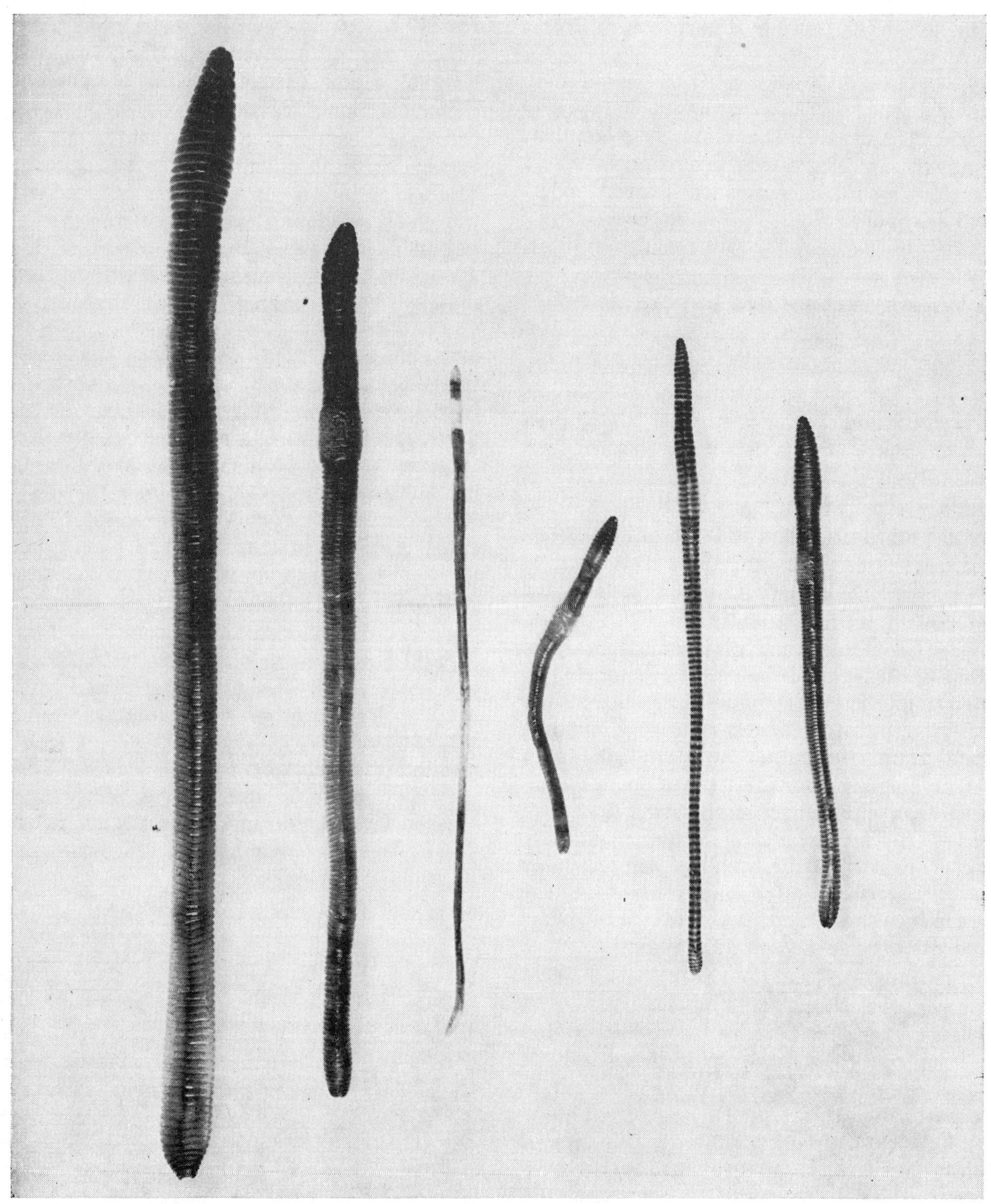

Figure 1

Some of the more common kinds of earthworms, from left to right: the night crawler, the field worm, the *Diplocardia* earthworm, the green worm, the manure worm, and the *rubellus* worm. The first four occur in farm or garden soil. The other two occur mostly in compost piles, but occasionally in garden soil to which a large quantity of refuse has been added. Natural size.

In addition to these recognized species of earthworms, there is a so-called "hybrid" that is being sold by commercial growers. It is claimed by some to be a cross between the manure worm and one of the field worms. This claim is groundless so far as we can determine. Shipments of these worms that we have examined proved to be identical taxonomically with the manure worm. Their rates of cast production are also similar.

Where Worms Are Important

All of the important earthworms are exotic, having been introduced from Europe, probably in soil brought along with plants. However, earthworms have become widely distributed throughout the country. By now, their distribution is largely a reflection of natural variations in climate and soil from place to place.

They are more prevalent in the humid sections of the East than in the arid West. But even in desert regions they sometimes occur along water courses and in irrigated land.

Where reasonable moisture conditions prevail, their occurrence is determined primarily by soil variations. They are more common in soils derived from limestone or otherwise rich in plant nutrients, than in shale or outwash soils. Soil texture influences the earthworm population. Sandy soil contains fewer earthworms than clay soil. This is fortunate because sandy soil is likely to have good structure naturally, whereas clay soil packs together and becomes too hard for crop growth unless agencies like earthworms are present to keep the soil granulated.

Another cause of variation in the distribution of earthworms is the organic matter of the soil. Earthworms require both organic debris and mineral soil for their food. So if there is little or no organic matter available in an area, earthworms will not be present. On the other hand, it is unnecessary to provide large amounts of organic matter to have earthworms; only a moderate quantity is necessary to support a reasonably large population if other conditions are favorable.

These variations in climate and soil result in very large differences in the size of the earthworm population. There are differences in the size of the population regionally, due to climate, and locally due to the texture and origin of the soil. There may even be differences in the population from one field to the next on the very same soil due to the method of cropping the land. Hence an actual examination of your soil is the surest way of knowing whether you are in a locality where earthworms are important.

Such an earthworm examination of your soil can be made quickly. Select a field that has ample vegetation on it, such as clover, grass, or alfalfa. The examination can be made at any season of the year, although easiest during the humid period of the year. Dig out a square of earth about one foot across to a depth of approximately seven inches. Count all the earthworms you can find in this sample. Include both mature and young ones. Our experience has given us a fairly reliable thumb rule for judging the earthworm population. If a soil contains at least ten earthworms in such a sample, the population is large enough to be significant in the structural properties of the soil. Sometimes there are only one or two earthworms in the sample. This indicates that the earthworms are playing little if any real part in the physical condition of the soil.

It was once thought that earthworms are confined to only the occasional pieces of rich ground. This is now known to be false. Of course, there are more earthworms in the richer soils. But even ordinary garden or farmland may contain a surprisingly large population. Some idea of the number of earthworms that occur in different sections and how the number can vary may be seen from Table 1. The values are fairly representative of ordinary soil where cover is present.

Table 1

Earthworms at various locations.

Location	Earthworms to a 7-inch depth Per Square Foot (Number)	Per Acre (Number)
Marcellus, N. Y.	38	1,600,000
Geneva, N. Y.	28	1,200,000
Ithaca, N. Y.	4	190,000
New Brunswick, N. J.	28	1,200,000
Frederick, Md.	50	2,200,000
Beltsville, Md.	8	350,000
Morgantown, W. Va.	28	1,200,000
Zanesville, Ohio	37	1,600,000
Coshocton, Ohio	5	220,000
Wooster, Ohio	30	1,300,000
Holgate, Ohio	14	600,000
Lansing, Mich.	13	570,000
Dixon Springs, Ill.	20	870,000
LaCrosse, Wisc.	39	1,700,000
Mayaquez, P. R.	6	260,000

What Worms Do For Soil

The newest researches have amply demonstrated that earthworms *do* have very important effects on the productivity of soil. They are by no means just passive denizens of the soil, without effect on its properties. Quite the contrary, controlled experiments now show that some of the important properties of certain kinds of soil are directly attributable to the activity of the earthworms, and that when the earthworms are absent these properties are altered.

Unfortunately, there has been a great deal of misunderstanding about what earthworms do for the soil. Just as there are some individuals who, oblivious to the facts, stick to the old idea that earthworms are without any real importance in farm soil, there are others who, over-enthusiastically, proclaim the earthworm a cure-all for every soil trouble. Both these extreme attitudes are wrong. There are many conditions that affect the productivity of soil, and earthworms change only certain of these conditions. A clear understanding of the role of earthworms requires, therefore, a clear understanding of what makes soil productive.

The ability of soil to produce bountiful crops depends primarily on (a) a supply of moisture in the soil, (b) an adequacy of air spaces for root development, (c) available nutrients. If any one of these three basic requirements is missing, crop growth will be poor, no matter how much of the others is present. However, no absolute values can be established for these factors, since crops vary in their minimum requirements.

The above requirements are, basically, all that have to be considered in the productivity of soil. That is why crops can be grown quite successfully in so-called sand or nutrient culture. With this method you use water, essential minerals, and a coarse medium, such as sand. Organic matter, earthworms, or any of the other multitudinous considerations that go into the tilling of ordinary soil are not necessary in sand cultures.

But ordinary land is no such ideal medium for crops. Natural soil is usually deficient in one or another of the three basic requirements and sometimes these deficiencies become more acute due to cropping. So, in most soils, to provide the three basic requirements, the farmer has to adopt corrective measures. Much of soil management science has to do with the discovery of measures that correct the deficiencies in the different kinds of soils where one or more of the basic requirements is limiting the crop growth. Thus, where moisture is limiting, the farmer may irrigate; or he may use a system that lets more of the rain into the ground so that not as much is lost by runoff; or he may increase the depth or capacity of the root zone to hold water. Where aeration is poor, such as in compact clay soils, he cultivates and may use other measures for increasing the volume of large holes in the root zone; or where poor aeration is due to an overabundance of water, he may bed or drain the land. Where sufficient nutrients are lacking, he uses mineral fertilizers; or he may grow legumes that bring nitrogen into the soil from the air; or he may supply missing nutrients in manure or other organic debris that he brings onto the land.

All these measures, and others too, are adopted by gardeners and farmers because soils, as they actually exist on our lands, are not perfect media for the particular crops grown. Yet no one of these corrective measures is absolutely indispensable. There are usually several alternative ways of accomplishing the same basic purpose. However, some of the measures are practical while others may be possible only theoretically.

Earthworm activity is not one of three basic requirements for plant growth. Rather, it comes into the category of factors which can be used to correct deficiencies in these basic requirements. Like any of the other corrective measures, earthworm activity is not indispensable. But, on some soils, it is difficult to find other practical means of doing those things which are normally done by the earthworms.

Effect on Soil Moisture

Soil that consists of coarse particles, such as sand, usually absorbs water readily. But most soil is not of this kind. Our principal kinds of agricultural soil consist of finer particles, such as silt and clay loam. Unless modified by secondary agencies, the particles pack together and become almost impervious to water. Then most of the rain runs off the surface instead of entering the soil. When soil gets into this condition, the crops suffer from drought even though the location may be one where there is adequate normal rainfall.

The intake of water by fine soil is dependent for the most part on the presence of extraneous channels. Earthworms are highly effective in making such channels. They form an interconnected web of channels which allow rain water to penetrate quickly throughout the topsoil layer. Figure 3 shows what happened to water in a silt loam soil with and without earthworms. The water found its way between the soil particles only slowly, but ran down the earthworm channels quickly. As a result, the soil without worms had an initial absorption rate of 0.2 inches of rainfall per minute. The same soil, worked by earthworms for a period of one month, had an

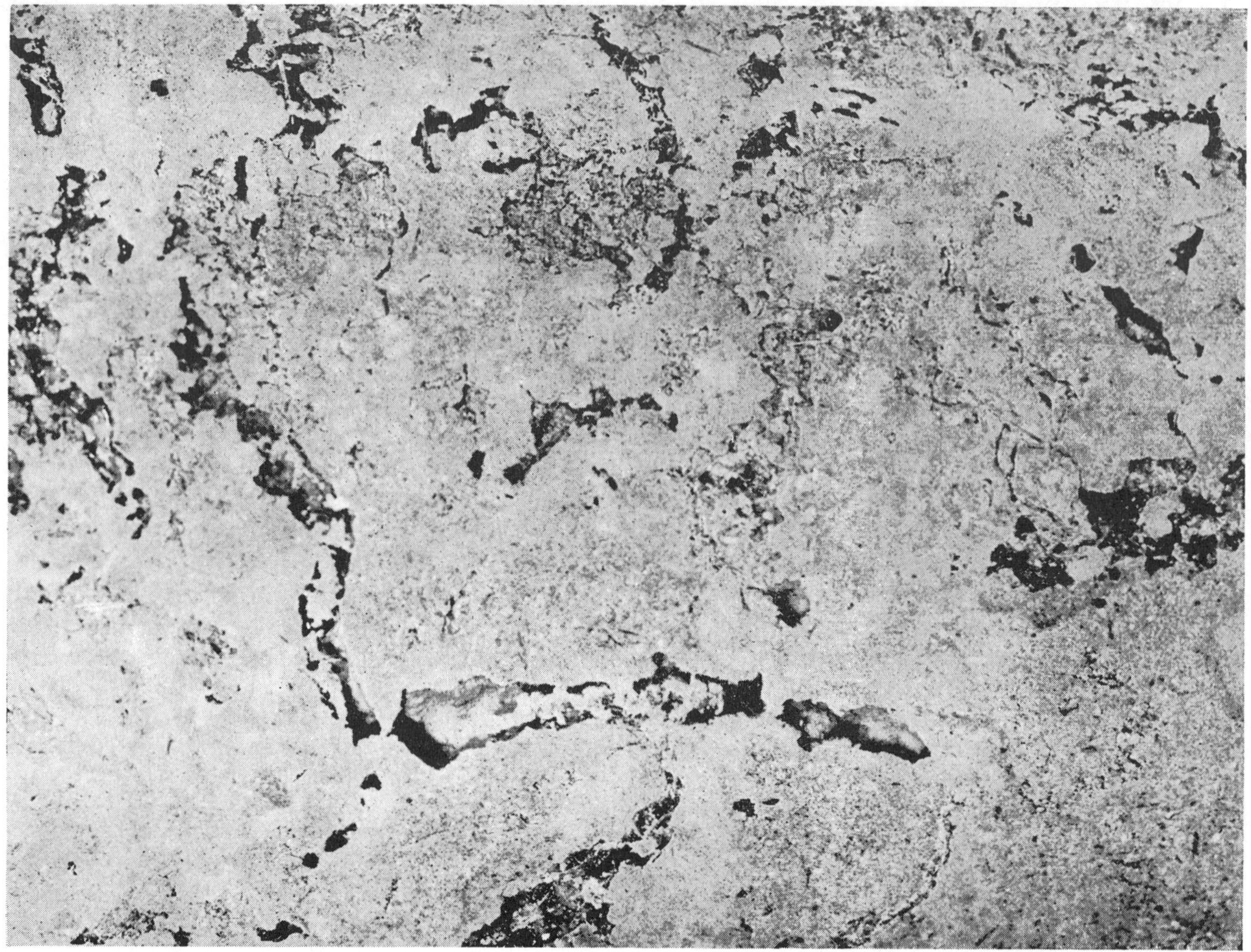

Figure 2

The interconnected web of channels that earthworms make are highly important for transmitting water in the soil. Here is a close-up of a fine-grained soil that has been put into good physical condition by earthworms. Actually, the channels interconnect to form a continuous network. Earthworms coat the walls of the channels with a secretion that makes the holes stay open even when full of water.

initial absorption rate of 0.9 inches of rainfall per minute, a fourfold improvement. The soil had about the same total capacity for rainwater, whether worms were present or not; but when earthworms were absent, the rate of intake was so slow that most of the water ran off.

The activity of earthworms is just one of several factors that increase the water-absorbing ability of soil. Mulch, compost and the roots of sod plants likewise increase infiltration. Table 2 shows how these various factors interacted to give the maximum rate of water absorption in an outdoor test on clay soil. In this test, sod and mulch were established both with and without earthworms. Two years after initiation of the test, infiltration measurements were made. Without any cover or earthworms, the soil remained in its original impervious condition. Addition of earthworms was without effect because the earthworms died for lack of food. Sod alone increased the infiltration rate somewhat. But when earthworms were added too, a really large improvement occurred. The soil then became highly permeable. The earthworms used the organic matter furnished by the sod and mulch for food, and by their activity made a network of channels that allowed the ready entry of water. These data indicate that the earthworms had much

Table 2

The influence of earthworms, sod, and mulch on the infiltration capacity of a clay soil.

Cover	Relative rate of infiltration	
	Without earthworms (inches/Minute)	With earthworms (inches/Minute)
None	0.0	0.0
Fertilized sod	0.2	0.8
Mulch	0.0	1.5

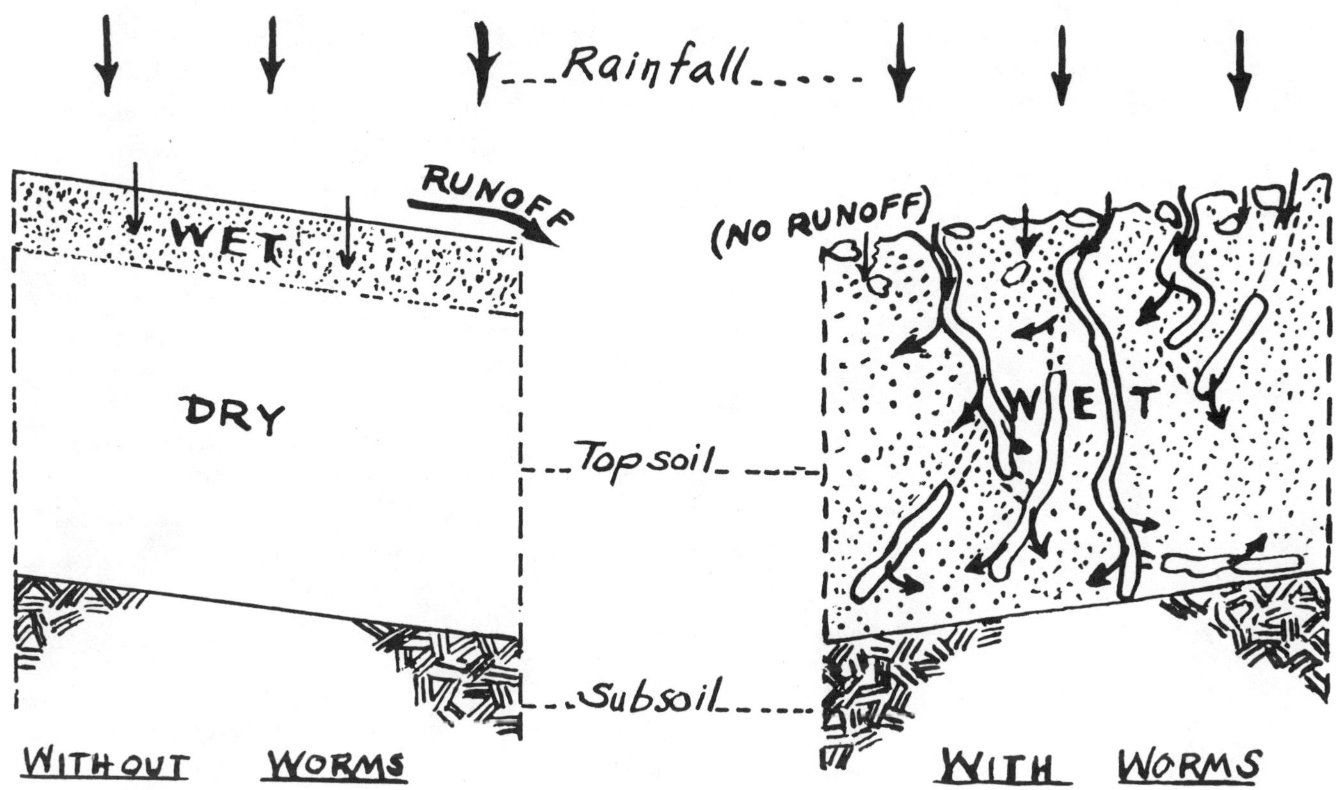

Figure 3

A diagram to show how earthworm channels improve water intake of fine-grained soil. In the absence of channels, the water wets the soil from the top downward. Often the rate of intake is so slow that much water is lost by runoff though only the upper part of the topsoil has become wet. When earthworms are present, the water runs down the channels and wets the whole topsoil evenly. The topsoil layer fills from the bottom upward. Runoff does not occur until the whole topsoil becomes wet.

more effect on infiltration than the sod or mulch alone.

When we consider all the different kinds of soil, we find that the infiltration rate depends on several different factors. So it is not possible to attribute the infiltration rate to any single factor. However, in general, there is a close association between the infiltration rate and the number of earthworms. Rarely does one find soil with a large earthworm population that does not take water fairly rapidly. Likewise, soil with few earthworms usually takes water slowly, though there are exceptions where some of the other factors besides earthworms are effective in keeping the soil open.

As might be expected, the number of earthworms in the soil follows rather closely the amount of runoff and erosion. Soil that does not erode retains its organic matter and this provides food for earthworms. Conversely, the more earthworms there are in the soil, the better the intake and the less the runoff. And erosion cannot occur unless there is runoff. Table 3 shows how the size of the earthworm population tends to follow rates of runoff and erosion. The association is not exact since other factors also enter into the picture. But the association is close enough to suggest that a high earthworm population is desirable in decreasing erosion and runoff.

Table 3

Earthworms, erosion, and runoff in comparable plots at Ithaca, N. Y.*

Treatment for the prior 10 years	Earthworms per acre (Thousands)	Erosion annually (Tons/acre)	Runoff annually (Inches/acre)
Continuous corn	0	30	1.77
Three-year rotation	94	5	.40
Continuous meadow	314	0	.18
Idle	813	0	.29

* Erosion and runoff data are 10-year averages; information supplied by Dr. John Lamb, Jr.

Effect on Aeration

You have probably noticed how poorly crops grow if the soil is compact. Growth is poor because there are not enough large air spaces in the soil. Not only is the absorption of rain water slow, resulting in its loss by runoff, but also

Figure 4

Earthworms gradually deepen the layer of topsoil and put it into productive condition for crop growth. In the lower part of the picture is clay subsoil. Earthworms have just started to penetrate it with channels, making holes, which are lined with their humus-rich casts. The soil in the upper part of the picture has been converted into productive, dark-brown topsoil and completely riddled by earthworms.

WHAT WORMS DO TO SOIL

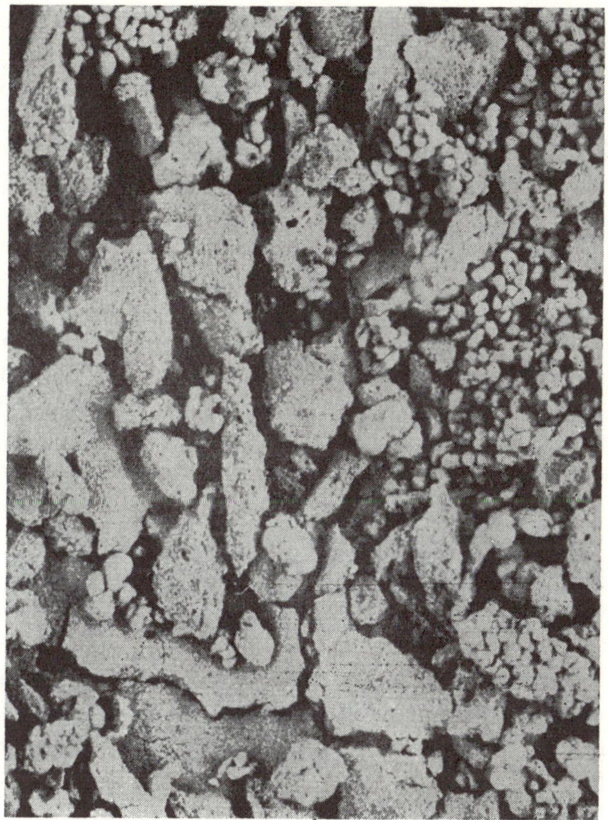

NO WORMS

WORMS

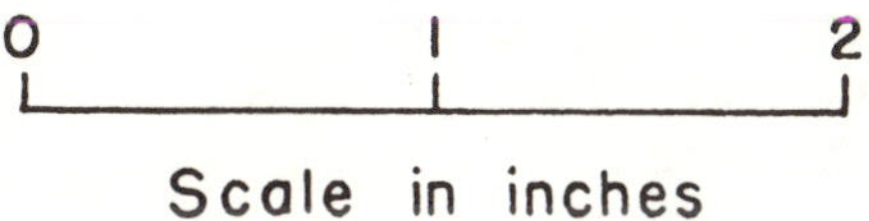

Figure 5

What worms do to soil. The soil at left without worms is compact; that at right with worms is granulated. Note the numerous oval casts in the ground. These are highly important for good structure in soil because they keep their identity when they become wet. Soil that is compact or that slumps into a compact condition after tillage is unproductive for crop growth.

aeration of the soil is deficient. Crops make their best growth where there are large spaces for the roots to grow in. Roots do not grow well in solid soil; they grow primarily in the spaces between the soil particles.

Earthworms are one of the most effective agents for loosening and aerating the soil. Their burrows make large passageways for the roots to grow in. They perforate the topsoil especially and gradually penetrate the subsoil, opening it for root growth and depositing organic matter in it (Figure 4). But even more important is the granulation of the soil which they bring about. This is done by their production of casts from the soil and organic debris that they eat. As the soil becomes granulated with casts, it gets looser and looser. These casts are clearly visible in any soil inhabited by earthworms (Figure 4). During damp seasons of the year, cast production is especially prolific. At that time, casts are even deposited on the surface of the ground. However, there are always many more casts underground than there are on top.

The casts are distributed for the most part in the topsoil layer. As a matter of fact, much of the dark, granular material so characteristic of topsoil often proves, on close examination, to be earthworm casts. The dark color is due to the admixture of organic matter, or humus, with the mineral material. This mixing process takes place inside the earthworms' bodies.

The rate of cast production depends mainly on the size of the earthworm. Large species produce a greater quantity of casts than small species; and mature ones a greater quantity than young ones. An approximate rule is that earthworms produce their own weight of casts per day. This rule is based on careful measurements in controlled tests. Transposed onto a field basis, and using data from a large number of examinations in the North Atlantic and North Central states, it appears that the average quantity of soil converted into casts amounts to about 700 lbs. per acre for each day's activity. This rate of activity holds in the damp periods of the year only; for earthworms become dormant as the soil becomes dry.

It has been mistakenly assumed that the loosening of soil by earthworms has the same kind of effect on soil as cultivation with tillage implements. Some people even think there is a controversy between earthworms and the plow. This is an unfortunate misconception. They do not do the same job. From the viewpoint of simple loosening soil, tillage implements are very much more effective than earthworms.

But if the soil is run down, it does not stay loose after tillage. With the first rains, the clods slump together and the loose condition made by tillage is lost. Thus, tillage gives the soil only a start with good structure (besides, of course, killing the weeds), but the ability of the soil to remain loose depends upon the properties of the soil itself. This property is known as water-stability.

Soil scientists know that the water-stability of soil is one of its most important physical characteristics. They have, therefore, given much attention to learning how soil acquires this property and they have developed precise tests to measure it. They now know that water-stability comes from the cementing of the soil particles together by sticky materials. These materials, once dried, do not redissolve in water. They are produced by the life in the soil, such as earthworms and certain microorganisms. It is in this respect that the type of granulation produced by earthworms differs so greatly from the type produced by tillage implements.

The effectiveness of earthworms for increasing the water-stability of soil is one of their most valuable attributes. This is shown by the data in Table 4, which gives the results of careful tests on various kinds of soil. The soils were incubated, both with and without earthworms, for periods of one week. Then the entire soil, not only the casts, was given the water-stability test. A consistent increase in water-stability was obtained in the soil containing earthworms as compared with the same soil incubated for the same length of time without earthworms.

You can easily see for yourself how earthworm casts increase the water-stability of soil. Pick up a few earthworm casts from a garden either from the surface or below ground. Then select a clod that does not consist of casts and break it into pieces of about the same size as the casts. Drop some of both in water. The earthworm casts stay whole for some time while the pieces of clod quickly break apart in the water. The difference in water-stability illustrates why earthworm casts help keep soil loose when it becomes wet.

The effect of this burrowing and granulating activity is reflected under field conditions in a

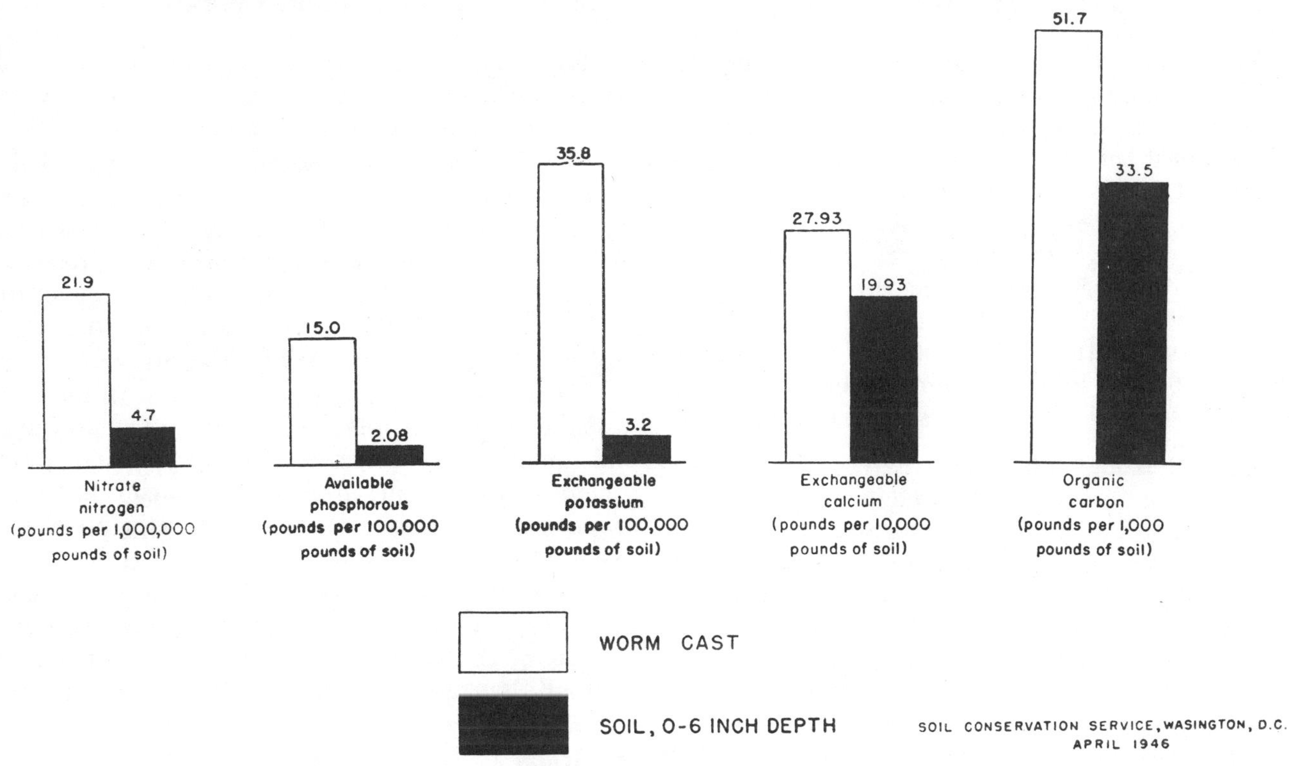

Figure 6

Chemical analysis of earthworm casts compared with the parent soil from a depth of 0 to 6 inches in a field. The casts in the soil are rich sources of chemical nutrients for crops. The extra nutrients in the casts came from organic debris that the earthworms took from the surface of the ground.

close association between the numbers of earthworms and the large spaces in the soil. Examples of the association are shown in the data of Table 5. These large spaces let water and air through them readily. In size, they are spaces like one finds between the grains of a coarse sand and larger. The meadow soils averaged 10 per cent of large spaces while the more compact bare land only 3 per cent. In each location, the large spaces followed the number of earthworms quite closely.

Table 4

The effect of earthworms in one week on the water-stability of soil taken from different places, as found in controlled laboratory tests.

Soil Type	Description of Soil	Water-Stability of the Soil: Without Earthworms (Per cent)	With Earthworms (Per cent)
Evesboro loamy sand	Eroded bank	2.5	**14.4**
	Same, with organic debris added	12.9	32.3
	Another eroded bank, with organic debris added	6.6	15.0
	Abandoned brushy field	18.2	45.5
Sunnyside fine sandy loam	Eroded field, with organic debris added	4.7	10.3
	Clover meadow	73.2	79.5
Leonardtown silt loam	Eroded bank	1.3	4.8
	Same, with organic debris added	4.2	8.4
	Young hardwood forest	74.8	80.4
Christiana silt loam	Lespedeza meadow	4.5	15.7
Paulding clay	Continuous corn field	19.8	24.5
	Rotation corn field	31.1	39.2
Average of all tests		21.2	30.8

Table 5

The association of earthworms with the amount of large spaces in the soil for land in various crops, measured in the spring of the year.

Location	Condition of Land	Earthworms per Sq. Ft. (Number)	Large Spaces in the Soil (Per cent)
College Pk., Md.	Bare, following corn	2	4.0
	Young wheat following corn	6	4.2
	Meadow following wheat	22	7.4
	Continuous meadow	26	12.6
Wooster, Ohio	Bare, following corn	2	1.0
	Young wheat following corn	8	2.8
	Meadow following wheat	14	4.4
	Continuous meadow	9	8.2
Holgate, Ohio	Bare, following corn	5	3.3
	Young wheat following corn	15	4.8
	Meadow following wheat	31	6.9
	Continuous meadow	38	10.7

Effect on Nutrients

Earthworms affect the nutrient supplying ability of soil by taking organic debris from the surface and incorporating it into the topsoil. They digest the debris and excrete in their casts what nutrients they do not need for their own nutrition. These casts are deposited in their channels throughout the topsoil and some even in the subsoil. So distributed through the root zone, the casts constitute a source of nutrients for the vegetation. The richness of the casts depends on the kind of organic debris and mineral soil that the earthworms have for food.

Very thorough chemical analyses have been made of earthworm casts and uneaten soil by H. A. Lunt and H. G. M. Jacobson of the Connecticut Agricultural Experiment Station (Figure 6). They report that the casts contained about 5 times the nitrate, 7 times the available phosphorus, 3 times the exchangeable magnesium, 11 times the potash, and 1½ times the lime (calcium) that occurred in uneaten soil from the top 6 inches of a field. The increases came from the organic debris that the earthworms ate.

Some people have suggested that earthworms add to the nutrient content of the soil. They think that earthworms free the chemicals that are in the mineral soil they eat. Actually, there is no good evidence to justify such an idea.

We have conducted laboratory experiments with the specific purpose of finding out if earthworms free the nutrients in mineral soil. Earthworms were placed in soil for 30 days. At the end of that time, the earthworms were removed and the soil analyzed chemically. The analysis showed no differences in the available mineral nutrients (phosphorus, potassium, and calcium) between the soil eaten by earthworms and the same kind of soil similarly treated but without earthworms.

However, there was an increase in the nitrogen content of the soil containing earthworms. But the nitrogen did not come from the mineral soil. Rather it came from the earthworms themselves. Some of them had died and others, while still alive, had lost weight. The increased nitrogen in the soil was accounted for by the decreased nitrogen in the bodies of the earthworms. We will see later how this nitrogen in the bodies of the earthworms can be important to crop growth.

Is It Necessary To Plant Earthworms?

Earthworms are already widely distributed over the country and it is rarely necessary to introduce them artificially in arable land.

If there are no earthworms at all in your soil, there is some condition that prevents them from living there. You should check into the moisture, kind of soil, supply of organic cover, and presence of toxic chemicals to find the cause. Planting of earthworms is meaningless when there are conditions that keep the earthworms out.

Where earthworms are present, but in too few number, artificial planting is not likely to have any notable effect either. Usually you need only modify the previous methods of treating the land in order to increase the population to an effective density.

On most farm and garden land, therefore, you can keep earthworm activity at a desirable level by using the correct soil management practices rather than introducing earthworms artificially.

There may be a few situations where it is desirable to raise earthworms for planting in soil or compost pits. Methods of raising earthworms for this purpose are similar to the methods used in raising earthworms for fish bait.

There has recently been considerable popular emphasis placed on introducing "hybrid" or specially-bred earthworms for soil improvement. To the best of our knowledge, there are no hybrid earthworms. Neither do we know of any artificially developed strains with special soil-building attributes not found amongst the various kinds of natural earthworms.

Building The Worm Population

Earthworms are very sensitive to changes in the soil brought about by cropping the land. In localities where earthworms are present, the population varies greatly according to the farming or gardening methods used.

Recent investigations have disclosed efficient methods of maintaining earthworms on tilled land. These new practices came out of studies on the life habits of earthworms.

It was found that earthworms follow a well-defined yearly cycle. One might consider the cycle as starting in the fall of the year (Figure 7). At that time, many of the earthworms are young, just starting their life. With the advent of wet, cool weather, they become extremely active physically. They feed on the organic debris

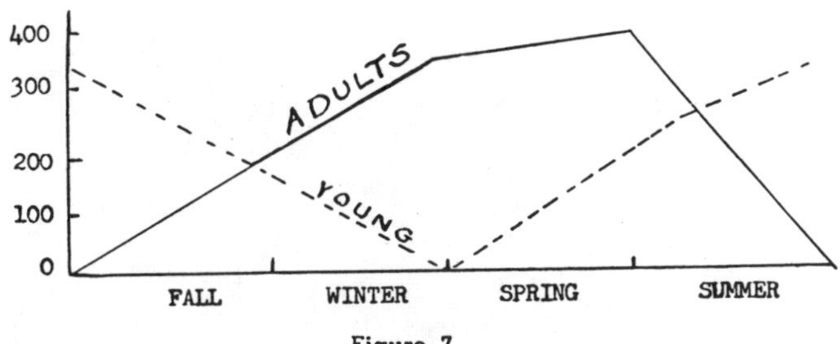

Figure 7

The annual cycle of earthworms in sodland. Diagrammatic representation of data from plots of the Maryland Agricultural Experiment Station.

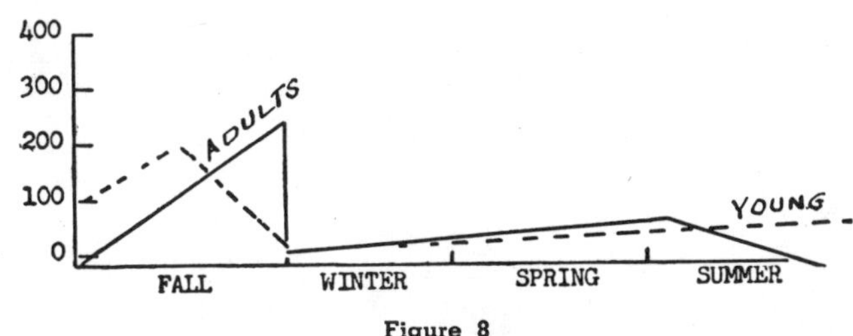

Figure 8

Earthworm cycle in a corn field. The earthworms are mostly killed in the late fall or early winter and they do not build back again until the following fall.

in and on top of the soil, and mix it with the mineral soil to produce casts, as well as making new burrows in the soil. During damp, cool nights, and occasionally on wet cloudy days, too, they come out of their burrows and seek new areas to inhabit. Within a few hours of a night, they may migrate a considerable number of feet. With dawn, they mostly disappear into the ground but the tracks they leave in the soft ground are evidence of their nocturnal meanderings.

The high level of physical activity normally continues throughout the fall, winter and spring. During this period, the young earthworms mature and more eggs are laid. During the winter, both mature and young earthworms, as well as eggs, can be found in the soil. But by late spring, most of the earthworms are mature. With the coming of summer, the soil dries and heats. The earthworms become less and less active. They lay eggs and many die. During the hottest and driest part of the summer, almost all the earthworms in a garden are young ones or unhatched eggs that had been deposited by the mature ones before they died.

Thus, summer is a period of sharp decline in physical activity of the earthworms. At that time of the year they have very little effect on soil. It is mainly a period during which the generation starts for the ensuing year.

This cycle is a reflection of the seasonal changes in weather. Differences in weather from year to year or from one region to another can modify the cycle somewhat. It can be modified also by keeping the soil moist and cool during the summer through watering or mulching. The earthworms will then be physically active throughout the year. However, the natural life-cycle fully adapts earthworms to the seasonal changes in weather without their requiring such artificial help.

The main reason that earthworms decrease on tilled land is the lack of protective cover and organic debris in the winter. Their physical activity is at a maximum during this time, and to carry on their activity, they require organic debris for food. If this is lacking, they die or leave the field in search of a more hospitable area.

In the northern section of the country, where the soil usually freezes during the winter, there is one other provision that must be made for the earthworms. They must be protected against freezing temperatures in the late fall. Earthworms change with the season in their resistance to cold. During the summer, they would die if they were subjected to freezing temperature. But by winter they are fully able to live in soil that is actually frozen. However, this change in tolerance develops only slowly. In the vicinity of Washington, D. C., the earthworms do not usually develop resistance to freezing temperatures until late December.

On land protected with sod or other debris, the soil temperature drops slowly as the cold weather of winter sets in. Bare ground, on the contrary, is subject to violent fluctuations in temperature. One or two days of low temperature in the late fall may freeze bare ground, especially at night, while soil under sod, protected by the insulative cover of leaves and litter, does not freeze. Under sod, then, the earthworms have sufficient time to adapt themselves to the freezing temperatures of winter, while on bare ground they may be killed before developing cold resistance.

An example of the drastic effect that freezing weather in the fall has on the earthworms in unprotected ground was observed about Washington in 1946. In two days of sub-freezing temperature at the start of December, two-thirds of the earthworms on tilled land were killed by freezing. This decline is shown in Figure 8 by

Figure 9

A good cover of leguminous sod established by late fall as in this sweet clover field, protects the earthworms and gives them food over winter. Their activity greatly improves the physical condition of the soil during the winter. The soil in this field took in water at a very slow rate, .01 inches per minute, in the fall but by the following spring, when this picture was taken, infiltration has increased to .11 inches per minute.

the sharp drop in mature worms at the start of the winter.

To protect the earthworms from freezing in the late fall, almost any kind of insulation is satisfactory, so long as it is not harmful to the earthworms. Manure, compost, chopped corn stover, wheat straw, a cover crop such as ryegrass which gives a heavy sod, or a mulch of dead grass and weeds make suitable protective coverings (Figures 9 and 10). Such materials are at the same time a source of organic food for the earthworms over winter.

You see that the unprotected soil is slumped. It has few large spaces for aeration and water intake. When dry, it gets hard. This condition is due to the destructive action of frost on the wet soil.

The soil worked by earthworms, on the contrary, is loose and granular. The numerous channels let water enter quickly and provide spaces for the roots to grow luxuriantly. Frost has not damaged this soil because the earthworm channels have drained out the excess water.

Soil can often be improved a great deal in just a single winter by protecting the earthworms. The degree of improvement attainable is shown by tests conducted in Ohio and Maryland. Land that had been cropped to corn in a three-year rotation was selected for the tests. In the late fall, after the corn had been harvested

Figure 10

This soil had been cropped continuously to corn for 7 years. Each spring it was reduced to a compact condition. The photo shows how the soil has been converted to a loose granular condition in a single winter by protecting the earthworms. The earthworm casts are clearly visible.

and the wheat had become established, some of the land was left uncovered and some was covered with a mulch. The following spring the condition of the soil and the earthworm population were measured.

The effect of the treatment was shown by comparing the covered land with the uncovered land and with adjacent sod land. Table 6 gives the data. Without protection over winter, the tilled land was in very much poorer condition than the sod. Covering the tilled land overcame much of the difference. The infiltration and earthworm population were as favorable as in sod, and the volume of large spaces was almost as good. Water-stability was still below that of sod land, but there was a substantial improvement from the treatment.

Even inert materials, such as boards and stones, will protect the earthworms over winter, as can be seen during the winter by looking beneath them where they are lying out on bare fields. Such things are, of course, not practical for covering tilled land, although they do demonstrate the wide range of materials that can serve as protection for earthworms during the winter.

These problems pertain more to farmland than intensively cultivated garden patches, where with good use of manure, green manure, and compost, there is little difficulty in maintaining

Table 6

Earthworms and soil physical properties in tilled land covered over winter compared with uncovered tilled land and adjacent sod land. Measurements made in the spring; average of 3 localities in Ohio and Maryland.

Soil Property	Tilled land in young wheat: Covered over winter	Uncovered over winter	Sod land
Relative infiltration (In./min.)	0.31	0.10	0.36
Large spaces (Per cent)	7.0	3.9	10.5
Water stability (Per cent)	59	44	79
Earthworms (No./sq.ft.)	52	10	24

Figure 11

In this experiment the soil in both barrels was fertilized, limed, manured, cultivated and seeded equally. The only difference in treatment was that the barrel on the right was inoculated with earthworms. The soil was a red clay loam in poor structure. The worms loosened the soil; the rate of water intake increased fourfold. The result was a luxurious growth of clover. Without worms (on the left) the same soil produced only a meager growth of poor-quality grass with little clover.

the earthworm population. But the dramatic differences which occur in farmland when protective practices are followed have both interest and significance for the small scale grower.

When organic debris and (in the northern states) insulative cover are given the earthworms during the winter, they prosper even if the land is tilled every year. Soil so protected is well granulated and porous the following spring, in contrast to unprotected land. An illustration of this effect is shown in Figure 10.

Laboratory feeding tests have shown that larger kinds of earthworms thrive best on organic debris that contains nitrogen, as in legumes, garbage, and manure. See Table 7. [These studies indicate also the value of adding nitrogen to compost material — Editor's note.]

Table 7

Change in body weight of the common field worm when fed different kinds of organic material together with a silt loam topsoil.

Kind of Organic Material	Change in Body Weight Gain (Per cent)	Loss (Per cent)
Lespedeza leaf litter	21.5	......
Fresh cow manure	10.7	......
Green clover leaves	7.8	......
Brown soybean leaves	5.6	......
Dead bromesedge leaves	......	0.5
Weathered corn leaves	......	5.2
None	......	13.4
Wheat straw	......	14.1
Sawdust	......	18.2

When the earthworm population is increased in sod land by good agronomic practices, marked changes often take place in the soil. As has been explained previously, earthworm activity alone does not explain all the changes that result, but it does influence some of the physical characteristics of the soil. At the Ohio Agricultural Experiment Station, for example, two areas of sod, consisting of alfalfa and timothy, were differently limed. One area was limed so that the soil was almost neutral (pH 7.0). The other area was limed only slightly, so that the soil stayed rather acid (pH 5.5). Measurements in the soil revealed the following:

Soil Measurements	pH 5.5	pH 7.0
Earthworm (No./sq.ft.)	4.3	13.5
Earthworm holes (No./sq.ft.)	22	58
Relative infiltration rate, (In./minute)	0.04	0.30

The holes referred to were those appearing on a horizontal cut made 3 inches deep in the topsoil. With the heavier liming, the earthworms increased, resulting in more holes in the topsoil. The greater number of holes allowed the water to go into the soil more quickly.

Effect on the Yield of Crops

Since earthworms improve several of the important properties of soil, it is only reasonable to expect better yields when earthworms are present. In general, this is true. A soil will usually yield more when a large earthworm population is present than when it has few earthworms. This is due in part to the effect of the earthworms themselves and in part to other productivity factors that parallel changes in the earthworm population. So the number of earthworms is a useful indication of the productivity of soil.

People frequently ask how much the yield of crops is increased by earthworms. A simple, direct answer cannot be given. It varies with the condition of the soil and the kind of crop.

This is illustrated by the following greenhouse experiment. Two crops, soybeans and wheat, were planted in a silt loam topsoil that was in good structure. The soil contained a large number of earthworms to begin with and they had granulated the soil rather well. These earthworms were removed at the start of the experiment. Also the soil was fertilized heavily. Thus, the soil was put in a highly productive condition, both physically and chemically. In a second set of plantings, the good structure was destroyed by puddling and compacting the soil. A third set was likewise puddled, but earthworms were then added to restore the structure.

The results of this test are given in Table 8. The soybeans were severely retarded in the soil with poor structure; growth was only one-fifth as much as in the soil with good structure. Earthworm activity restored almost all the difference. The wheat did not suffer as much from poor structure; growth was three-quarters of that in the soil with good structure. Again, earthworms restored the productivity.

Table 8

An experiment on the importance of earthworm activity to the productivity of a silt loam soil.

Soil and Earthworm Treatment	Weight of crop plant Soybeans (grams)	Wheat (grams)
Natural soil structure	2.96	9.5
Soil structure destroyed	0.56	7.1
Soil structure destroyed, living earthworms added	2.30	10.5

This experiment shows that earthworm activity becomes more important as the structure of the soil declines. Also, earthworm activity is more important with a crop, like soybeans, that is sensitive to soil structure. Most vegetables are very sensitive in this respect.

Figure 12

In this crock experiment, a soil with fairly good structure was used. It was fertilized, cultivated, and then treated as follows: (left) no earthworms; (center) dead earthworms mixed into soil; (right) living earthworms introduced. The dead earthworms resulted in three times the yield, the living earthworms in only slightly more.

When the structure of a soil is very bad, the improvement in plant growth due to earthworms can be little short of amazing. Figure 11 shows the effect of earthworms on the yield of hay in a red, clay loam subsoil. Both barrels were heavily fertilized, manured, limed, cultivated, and seeded. The treatments were the same except earthworms were added to the one barrel. Without earthworms there was a weak growth of vegetation, mainly grass and weeds. The average yield was at the rate of 0.6 tons of hay per acre. With earthworms, there was a luxuriant growth of ladino clover, averaging at the rate of 2.0 tons of hay per acre. The earthworms resulted in more than three times the yield, as well as a greater proportion of the highly desirable clover. Such a large amount of fine hay was particularly surprising for a soil of such low natural productivity.

In soils that have good structure — high porosity and water stability — no yield benefits can be expected from the physical activity of earthworms. The benefit of earthworms is then confined to their chemical effects on soil. Where a large earthworm population exists in soil, the release of their body chemicals during the summer, when the mature earthworms die and disintegrate, must be considered. The effect of the chemical content of earthworms on yield was determined in greenhouse test with lima beans and millet. A silt loam soil having fairly good structure was used. To begin with, all the earthworms were taken out of the soil, which was then heavily fertilized. Dead earthworms were mixed into the soil of one set, and living earthworms were put into another set. The yields are given in Table 9.

Table 9

An experiment on the importance of the chemicals in the bodies of earthworms to the productivity of a silt loam in good physical condition.

Earthworm treatment	Weight of crop plant	
	Lima beans (grams)	Millet (grams)
None	5.8	2.1
Dead earthworms added	16.0	2.7
Living earthworms added	16.9	3.0

Here, where the physical condition of the soil was good, the dead earthworms increased yield almost as much as the living earthworms (Figure 12).

These experiments lead to the following conclusions:

(1) Earthworms help the physical condition of soil by their winter activity. With poor-structured soil and crops sensitive to structure, an increase in yield can be anticipated. But little or no increase in yield can result if (a) the soil has good structure naturally, (b) you are growing crops that are not especially sensitive to poor structure, (c) growth is limited by a lack of other things such as fertilizer or water.

(2) Earthworms help the fertility of the soil during the summer by the chemicals that are released in the bodies of the mature ones as they die. With soils of low fertility and with crops having high fertility requirements, an increase in yield can be anticipated. But little or no increase can result if (a) the soil is naturally rich, especially in organic matter, (b) you are growing a low fertility crop, (c) the earthworms were killed out the previous winter by improper management practices, (d) growth is limited by a lack of other things, such as water or aeration.

Finding and Growing Worms for Compost and Bait

WE asked Henry Hopp specifically about the value of worms in the type of use with which we were experimenting, and his reply follows:

"There is no doubt that the species *Helodrilus caliginosus* (the common field worm) is of real value in helping break down manure, garbage and other vegetable matter. In one of my publications we showed data in which we studied the effects of having organic matter and soil with and without earthworms. The formation of water-stable aggregates was much increased by the presence of the earthworms. We observed that when organic matter and soil were placed in a container without earthworms being present, the organic matter gradually decomposed but did not become mixed intimately with the soil. After a time we could still find the decomposed bits of organic matter while the little areas of soil between the organic matter still retained the original color of the raw soil.

"Any container, whether indoors or outdoors, that is quite isolated from surrounding earthworms probably would be worthwhile inoculating. But if it is located in a spot where earthworms can get to it, this is not necessary. Whether one needs to plant earthworms artificially or just let them come in by themselves can be determined rather quickly by observation. If you put out a pile of organic matter and earthworms are able to get to it, in a matter of just a few weeks you will find some present."

Assuming you find that worms do not develop in your garbage pit, or that you want a large supply, say for fishing or for sale as bait, the way to find and cultivate them is also made available by Dr. Hopp, in a mimeographed bulletin of the USDA, from which the following is adapted.

Methods of Collecting Earthworms

The burrows of earthworms are often well concealed and, even where numerous, may not be noticed by the uninitiated. The worms have the habit of plugging the entrance to their tunnels with various materials, such as leaves, seeds, twigs, or pebbles, which remain until the inhabitants emerge.

During the more humid portions of the year earthworms ascend to the surface of the soil in the evening, to feed and to mate, and very probably for purposes of migration. That they emerge in great numbers during rains at night and travel considerable distances is a fact that is obvious.

Although the common way of obtaining earthworms for angling purposes is by recourse to the shovel, garden fork, or mattack, this is by no means the easiest or most effective method of collecting them. They may be more readily gathered by means of a lantern or flashlight, after dark, on almost any piece of well-fertilized sod, especially where there are a few large trees to furnish both leaves for food and shade from the summer sun. If a flashlight is used, it is well to subdue the light by covering the lens with tissue paper or a thin handkerchief, as the worms instantly withdraw into their burrows when too bright a light is flashed suddenly upon them, and not many creatures can move more quickly than a frightened earthworm. The best light for this purpose is fitted with a red glass, as red light will not alarm the worms.

In dry weather earthworms descend more deeply into the soil, and come to the surface in smaller numbers at night. At such times, indeed, they may seem to be entirely absent. Usually, however, it will be found that if the soil around their burrows is thoroughly wetted down by sprinkling with a garden hose shortly before sundown, the worms will come eagerly to the surface after dark, provided the night temperature does not fall much below 40 degrees. In case it is desired to obtain earthworms during cold weather, they may usually be found by digging beneath compost heaps or manure piles. Earthworms are hardy creatures, and a brief period of mild weather brings them to the surface even in February, in the latitude of Washington, D. C.

The worms should be collected in a clean vessel, such as a fruit jar or a thoroughly cleansed tin can. They should then be placed in a cool place, preferably in an electric refrigerator.

They should not, however, be allowed to remain indefinitely in a metal container, as they survive much longer and in better condition when kept in an earthen crock or a tight wooden pail or box. A 3-gallon stoneware crock, containing about 8 inches of mellow earth covered with a 2-inch layer of plant material such as sphagnum moss, sods, or even common chickweed is excellent as a storehouse for 100 or more large earthworms.

In such a container, kept in a cool corner of the cellar, the worms remain in good condition for weeks; almost indefinitely if fed a little finely chopped raw beef suet, crumbled hard-boiled egg, or even finely divided bread crumbs. These may be sprinkled on the surface of the soil in the container, underneath the plant material.

Storing and Rearing Earthworms

Where it is desired to store or rear earthworms for sale, a larger container placed out-of-doors is desirable. For this purpose a tight box, preferably constructed of tongue-and-groove material, is suitable. It should be at least 18 inches deep and if the other dimensions are 36 by 60 inches it will serve for several hundred large worms.

The lid should be well fitting, with a projection to prevent flooding in heavy rains. It should be set into the soil with the upper 2 or 3 inches projecting above the surface, in a fairly well-drained place, and should be shaded to prevent the temperature of the interior from rising too high in midsummer. A temperature of 75 degrees or higher is quickly fatal to earthworms under most conditions. The box should be nearly filled with good soil which is damp but not wet. The richer this soil in organic material the better; as we have seen, worms like garbage. A very sandy soil should be avoided.

After the box has been stocked with worms, the surface of the soil may be covered with a layer of cut sods if desired, but a very excellent covering consists of well-decayed leaves or lawn clippings. In dry weather it will be necessary to moisten the soil in the box occasionally, but in doing so care should be taken to avoid flooding it, as too much water is injurious to the worms. In severe climates, prevent freezing which would kill the worms by piling a generous covering of manure, compost or other such material on the box.

Although under such conditions earthworms can live for a long time without artificial aid, if there is little or no garbage going into the box it will be well to add a little protein fat and sugar in cheap forms.

Preparation of Earthworms for Angling or Market

Although earthworms may be marketed or used freshly dug from the ground, they are much more desirable, will live longer on the hook, and will take more fish if well "scoured" before use. This fact is well known to all skilled bait fishermen, and it is probable that the knowing ones would be willing to pay a premium for such worms. This scouring process has been known for hundreds of years and was well described by Izaak Walton in 1653.

"To scour," a quantity of spagnum moss such as is used by nurserymen in packing plants for shipment is put into a stoneware crock or tight wooden box. This moss, which grows in shady, swampy woods, should be well moistened, but the excess water should be wrung out before the moss is placed in the container. The worms should be placed in the moss for at least two days, and preferably three or four, and kept in a cool place. At the end of this period they should be almost transparent, tough and lively. In case it becomes necessary to keep them in the moss for some weeks, a little sweet milk should be poured over them at intervals of about a week, but the moss should be washed and wrung out in clean water every week or 10 days.

"Cultivated" Earthworms as a Business

For complete coverage of the earthworm subject, we wanted more information on the kind that are sold to gardeners, a business of quite considerable size. This section is by Carolyn Robinson and is told as a story of two city women who do earthworm farming and entertain summer guests at West Brooksville, Me. It is adapted from a chapter in "The Countryman's Guide," published by The Country Bookstore, Noroton, Conn.

You will find reaffirmed the twin truisms that worms do like garbage and people like worms!

GLADYS PEDERSEN and Minerva Cutler, dreaming like so many busy city folks of a farm or business in the country, as a semi-retirement project, decided to start "retiring" while they were still young.

They acquired a grand old 200 acre coastal farm at West Brooksville, Me., and started fixing up the run-down house for a children's camp. But the parents clamored to be taken too, so a business in summer guests was established.

That was in 1945, and it was also the year that Miss Cutler first heard about earthworms.

At a convention she heard about another teacher making $2,000 a year breeding worms in her spare time. The earthworms were in great demand because they produced wonderfully rich black compost. This immediately interested Miss Cutler when she remembered her discouragement over getting broken-off tops when she was trying to pull carrots out of her garden with soil so heavy with clay it was like cement.

Since we all still like to see to believe, Miss Pedersen didn't get too excited about the idea until she, too, heard firsthand about worms and their remarkable results. But it wasn't until the spring of 1948 that they bought 5,000 worms and turned them out to pasture in their compost heap.

Worms Are Real Multipliers

A year later their 5,000 earthworms had multiplied to an estimated five million! They were in business. And by 1952 with worm-eaten compost and mulch their garden was a thing of beauty despite the severe drought.

Now that Miss Pedersen and Miss Cutler are well established in the earthworm business, they know that their product can create as heated an argument as a presidential election. There are many people who believe that earthworms are oversold, but the real argument seems to boil down to: "Is there such a thing as a hybrid worm?"

The worms sold today are usually called "hybrid earthworms" and are said to have been bred by George S. Oliver, a Los Angeles physician. He claimed to have crossed a red manure worm and the deep-burrowing orchard worm to produce a desirable hybrid with the best characteristics of all natural worms. And while T. J. Barrett has carried on with the promotion of "hybrid worms" to the extent of riding into "Who's Who" on his worms, the few available research reports from the U. S. Dept. of Agriculture and other state extension services state that no one, as far as they determine, has done any serious breeding of worms and that the problem is so complicated that cross-breeding is probably impossible. It is also said that the so-called "hybrid worm" is really a manure worm and not a kind that thrives in ordinary soil.

Miss Cutler and Miss Pedersen have called their worms "hybrids" as they assumed Oliver was a reliable scientist, but as they say: "we sometimes wonder who cares where the earthworm came from and whether he is a hybrid. The important point is that our soil has increased tremendously in fertility through our use of them in our compost heaps and that is what concerns us."

In a report highly laudatory of earthworms and the nutrient values of their castings, the U. of Conn. asks: "Why spend money for worms?" — says "the addition of organic matter is the most satisfactory way of bettering the structure of the soil." Also, use of sufficient amounts of unsterilized manure would include large quantities of earthworms which would add castings to the soil and therefore "the mere addition of manure accomplished the same results as did the addition of castings."

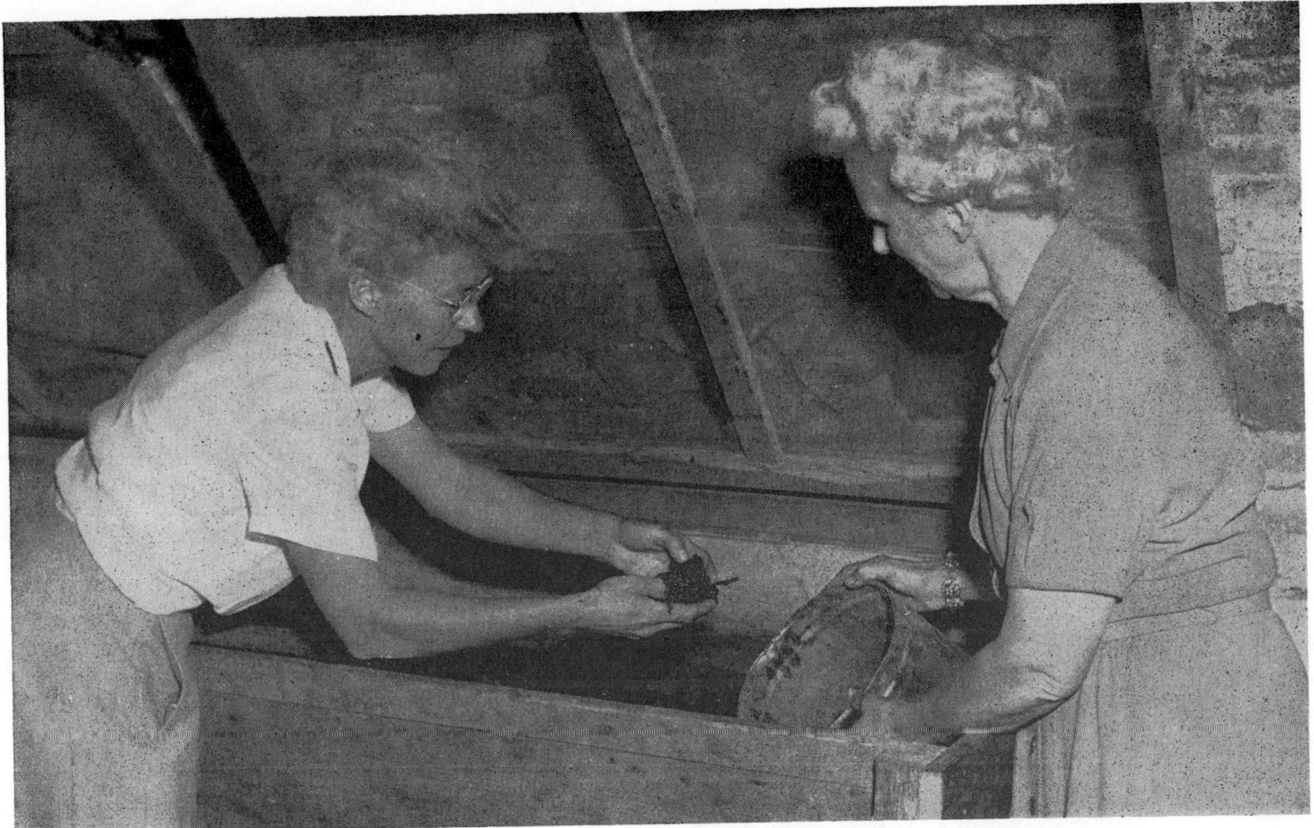

In this basement pit, the worms make their winter residence. A pit on legs makes work easier, but wooden bottom has to be replaced every two or three years. The screen doors are to prevent cats from disturbing the worms.

This last point is probably true, but most people interested in gardening of various kinds do not have access to "sufficient amounts of manure."

So "why buy earthworms?" Adding together statements of believers and non-believers, it seems to me there are sound reasons for doing so. Any one in a hurry to build up soil for a garden, shrubbery, trees or for any reason would find worms helpful in quickly decomposing whatever organic matter he can locate. If the organic matter is to be used in compost piles, the worms are especially helpful in turning the stuff over and homogenizing it. This matter of labor is the biggest reason why the compost pile has gone out of favor in place of sheet composting. But sheet composting is not always practical nor is it desirable for appearance sake in the case of lawns, flowers or shrubbery. When it comes to converting everyday garbage into usable matter, I have personally found a worm pit in the basement in winter or an outside covered bin in summer the easiest, cleanest and quickest method. It beats an outside compost pile which is as irresistible as Chanel No. 5 to dogs and skunks and which is not easy, sometimes impossible, to handle in winter. The worm pit is easier than burying garbage directly in the garden.

Buying is simplest

Granting that a person becomes interested in using worms for a compost pile or a garbage pit, it is easier to buy worms than to dig for them, just as it is easier to set in started tomato plants than it is to plant seeds. Also no one seems clear as to whether the ordinary garden worm does well in a garbage pit — worm enthusiasts say not.

If you become interested enough in earthworms to consider them as part-time business, there is a wide, sometimes repeating market despite the argument. Worms are not only used by gardeners and houseplant growers, but also by bird breeders, fishermen and some poultrymen, for food and as deodorants in pits under roosts.

The repeating fish bait business is said to be particularly good. One man, A. L. Dunn, of San Leandro, California, started selling worms for a full time job. He produces more than 10 million

Here it is in black and white. At left, original clay garden soil—at right, same soil after worms have worked on it mixed with organic matter.

worms a year in glasshouses and outdoor beds on one acre of land.

However, most people grow earthworms on a part-time basis and start out with a minimum of outlay, as did Misses Pedersen and Cutler. Their business grew so rapidly in one year that Miss Cutler gave up her teaching so that she and her partner could devote full time to the worm business and the summer guests. These two businesses fit together like the wheels of a clock but notice that their transfer from the city was done gradually . . . with "one foot in town — one foot on the land" until success was almost certain.

As for the investment necessary, it varies entirely with the individual and what he or she wants to do. The initial investment in worms depends on how soon you wish to start selling. From this farm you can buy 1,500 worms in all stages of development for $10, 3,500 for $25 or 10,000 for $50. The $50 order furnishes the worms at ½¢ each, but the $25 order will get you off to a start. These two partners recommend worms in all stages of growth as young ones adapt themselves quickly to change.

For a pit they suggest one 4 by 8 by 2 feet as this allows for easy access across it and into it. It must be drained and naturally concrete blocks or cement would be most durable. Any old boards can be used, but the bottom has to be replaced every two years or so. Some growers raise worms in stacks of vegetable lug boxes costing 10¢ each, but this method involves more shifting around of heavy weights. Worms can also be grown outdoors in mild weather in any kind of a pit or compost box but have to be protected from animals.

For the finest bedding material "the worm ladies," as some people call them, recommend rotted manure or compost, peat moss and soil mixed equally. Of these three items, only one has a definite price: peat moss costs the ladies $5 to $6 a bale and they use one bale per year. The rest of the feeding material is also highly variable and should be obtainable free — any meat, fish or vegetable waste from the markets is usable. If you run out of free stuff, chicken mash can be used.

Other expenses of the worm business also vary according to the individual. Misses Cutler and Pedersen use icecream cartons for shipping the worms: quart size for 250-300 worms, gallon size for 1500-2000 at 4¢ and 15¢ respectively. The cans and labels for packaging the worm castings can vary — for mail order they can be simple packages; for counter sale they should be sturdy cans with attractive labels.

Unless you can develop steady, repeating customers for your worms, advertising is the biggest expense, and obviously the more business you want the more advertising is required. One mail order expert says that a good return on a $25 ad would be $50 in sales. There would be little profit directly from the ad. That would come from the repeat orders.

Since Misses Cutler and Pedersen are only interested in selling worms for gardening, they find they do best in gardening magazines such as *Horticulture* and *Organic Gardening,* plus the special gardening sections of New York and Boston newspapers. Advertising in *Yankee* and *Popular Mechanics* has also proven successful. For the sale of worms for fish bait, they would use *Outdoor Life* and *Field and Stream.*

Advertising for Customers

The worm ladies have found the following advertisement sufficient to bring them in all the business they can handle:

"David's Folly" Earthworms will build topsoil for your garden, lawns and trees. "David's Folly" Earthworm Farm, West Brooksvale, Me. Free folder.

"David's Folly" was the name of the farm the worm ladies bought — they recognized it as a name with high remembrance value and have never changed it, even though the original owner raised cattle!

To fill their orders for worms the girls count their products, worm by worm, but wish for some kind of magnet. Most breeders sell by estimate.

Recently, the worm ladies have also started selling earthworm castings — one quart of castings for $1 plus 25¢ postage.

Preparation of earthworm castings for sale is quite simple. After the worm pit has been thoroughly worked through and the table scraps or other food has been dissolved, the black compost is removed into large metal cans (25¢ each from the bakery) along with a large quantity of worms. The worms remain here until they have eaten their way through and through so that the can contains pure castings. This can be determined by feel as the finished castings are like fine granules. The worms can be easily removed by placing a handful of fresh food in one place where they will collect within 24 hours. Then the castings are removed and dried in an oven at 200° for one-half hour.

David's Folly Earthworm Farm, West Brooksville, Me., has a scenic setting.

Miss Cutler and Miss Pedersen love their business — the worms as well as guests — and enjoy talking about it with other gardeners.

Both women have found a well-rounded, satisfying life by helping build a better community, as well as building better earth with the help of their handyman, the earthworm.

Role of Microorganisms in Garbage Composting

NOT only earthworms but microscopic life of many kinds operates on garbage to break down the organic matter and turn it into humus.

These organisms, for the technically inclined, include bacteria and fungi; the intermediate forms known as actinomycetes, protozoa (primitive animal life) and tiny insects. They will attack almost anything and can reduce to compost such stuff as citrus rinds and bones which are too resistant for earthworm action.

We could get along without the worms, but not without the microorganisms. However, the latter do not operate as efficiently without the worms. As Dr. Hopp pointed out, organic material reduced without earthworm help is not mixed satisfactorily, but remains in variable aggregations. Furthermore, the microorganisms we are interested in are aerobic, that is, require air as part of their needs for living, as compared with the putrefactive anaerobic organisms which operate without air. Earthworms keep the pit material open to air action, and eliminate need for turning over the stuff.

Generally speaking, the microorganisms needed for garbage disposal are found in ordinary soil. The microscopic life involved is fabulous in quantity. It has been estimated that in an acre of good soil there is a weight of microorganisms equal to that of a good sized cow. What these tiny forms do in the soil is approximately what they do in the compost pit.

Recent work on commercial scale with rapid composting of garbage by bacterial action opens possibilities which may yet be realized on a scale and with equipment and facilities available to the average gardener.

These new developments were initiated by Dr. Ehrenfried E. Pfeiffer, a refugee Dutch biochemist, and have been described in *Collier's* (May 31, 1952) and *Reader's Digest* (September, 1952).

Applying himself to the problem of how the decomposition of garbage could be speeded up, Dr. Pfeiffer recalled that European farmers believe that such plants as nettle, dandelion and valerian added to manure or compost will bring about more rapid reduction to humus. Are special strains of microorganisms harbored by such plants — or perhaps does a hormone stimulate growth and activity of such organisms already in the soil?

Pfeiffer went on to study all the strains of soil bacteria and other types of soil life that he could find. The net of several years of research, as the account goes, was the isolation of more than 50 strains of bacteria, some more effective on certain types of material than others; some operating best in summer, others in winter. And so on. The organisms were harmless to animals and human beings, but their appetite for garbage was phenomenal.

Pfeiffer's discoveries have been put to work on garbage collected from the city of Oakland, Cal. A conveyor belt carries the refuse brought in by the garbage trucks. A suction fan removes loose paper; a magnet plucks out metal objects. Then the garbage is ground and water added, in which Pfeiffer's cultures are mixed, about a tablespoon to a ton of garbage.

The pile is accumulated in the open. A furious action sets in, according to the description, and temperatures go up to 150 degrees, accompanied by huge clouds of steam. In about a week the action dies down, the pile cools off and what is left is the fine black crumbly stuff we get in our garbage pit — though not so fast. Maybe the English have something in the "activators" they use for composting!

Furthermore, it is said that Pfeiffer's cultures, about two ounces to an acre sprinkled in water, greatly speed sheet composting, enabling farmers more quickly to plant fields in which crop residues or green manures have been plowed or disced under.

From whatever the source, we seem on the threshold of an era in which the soil will be better replenished and the shameful waste of garbage and other good organic matter put an end to, or at least much reduced.

Do your part to help the worms and microscopic life, and enjoy better fruits, vegetables and flowers as well as the satisfaction of doing your bit for soil conservation, one of the most urgent needs of our time!